BEGINNING GRAPHING

Written and Illustrated by
Eleanor Villalpando

ISBN# 1-56175-445-5

©Remedia Publications. All rights reserved. Printed in the United States of America. The purchase of this unit entitles the individual teacher to reproduce copies for classroom use only. The reproduction of any part for an entire school or school system or for commercial use is strictly prohibited.

REMEDIA PUBLICATIONS   **10135 E. VIA LINDA, #D124**   **SCOTTSDALE, AZ 85258**

Dear Parents,

Our class has been learning about graphs. We have talked about how and why graphs are used. We have also learned how to read a graph to get information and answer questions. We have explored several different kinds and looked for ways in which they are alike. Learning how to collect information and show it in different ways has been an ongoing activity.

Graphing is an important skill being used more and more in all areas of the curriculum. It is very helpful for children to understand and use different types of graphs effectively.

As your child brings home graphing activities, please take a few minutes to talk with him/her about his/her work. Your questions will reinforce the learning and understanding of this important tool.

We will have a few home assignments in making graphs. I hope they will be considered a family activity in which you all can participate.

Thank you for your help and cooperation.

Name _____   Tallying

Count tally marks. Write the number.

 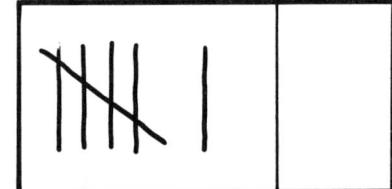

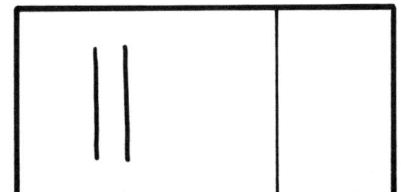 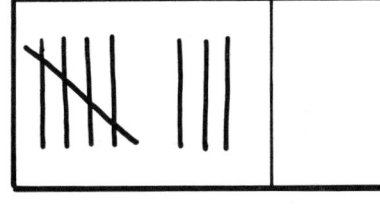

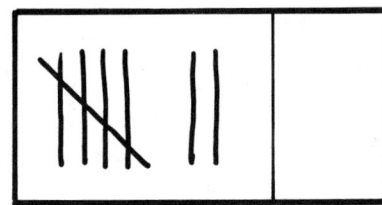 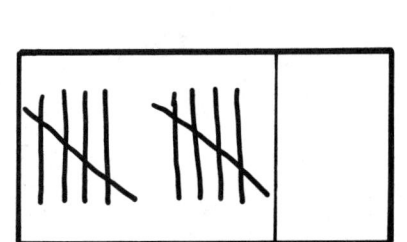

===

Make tally marks to show the number.

| 6 | | 8 | | 4 | |

| 3 | | 10 | | 7 | |

 | 5 | | | 9 | |

1

1989 REMEDIA PUBLICATIONS

Name _____   Tallying

Make a tally mark for each animal. Color the animals.

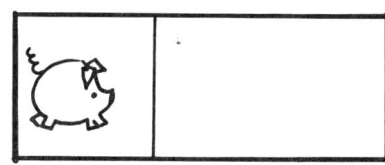

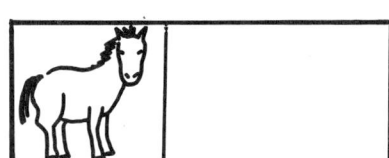

Name _____   **Tallying**

Make a tally mark for each goodie. Color each goodie.

Bakery

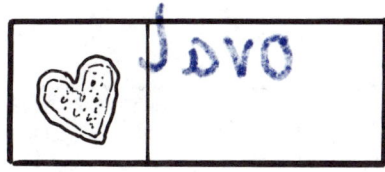 oval  circle  triangle

 rhombus  square hexagon

Name _____   Tallying

Make a tally mark for each shape. Color the shapes.

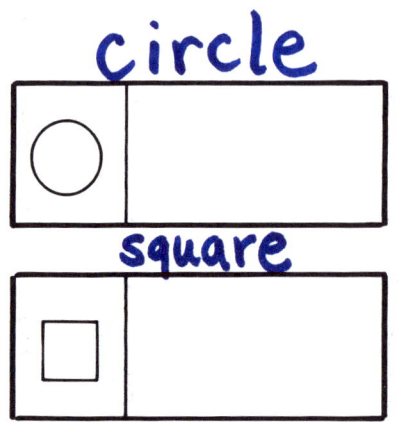
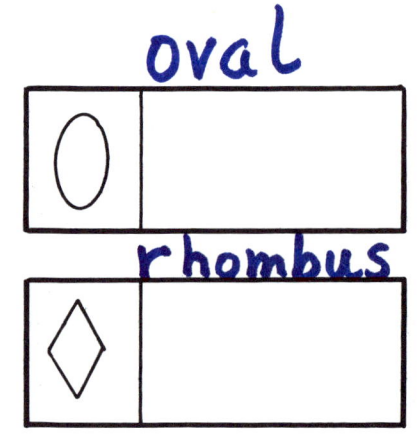

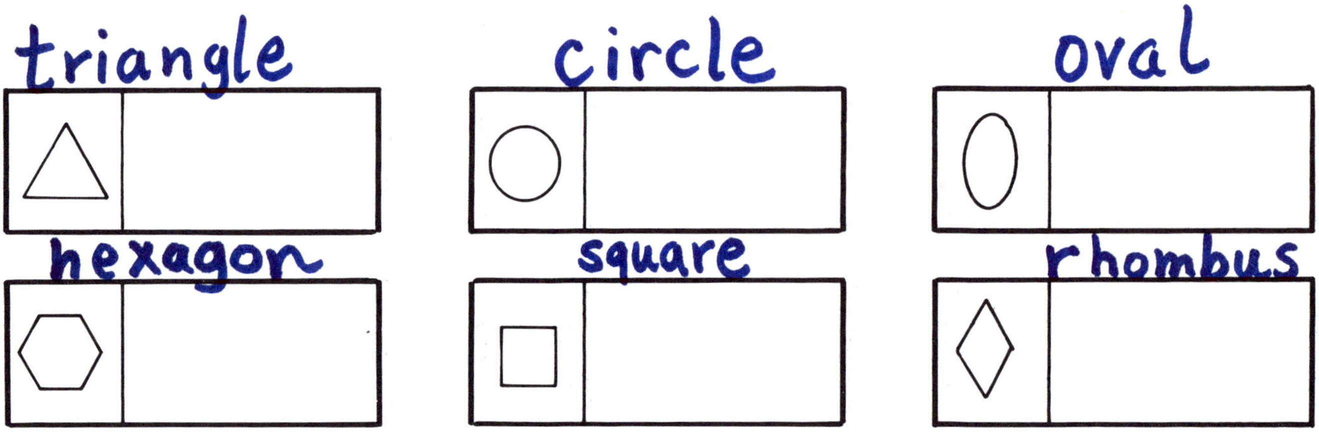

4

1989 REMEDIA PUBLICATIONS

Name _____

Tallying

Make a tally mark for each bug. Color the bugs.

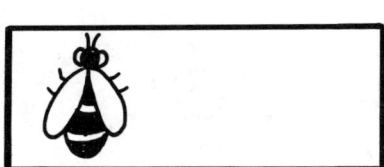
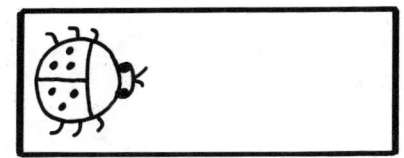
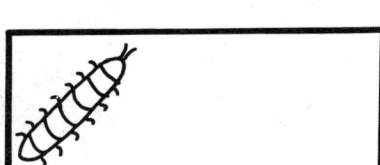

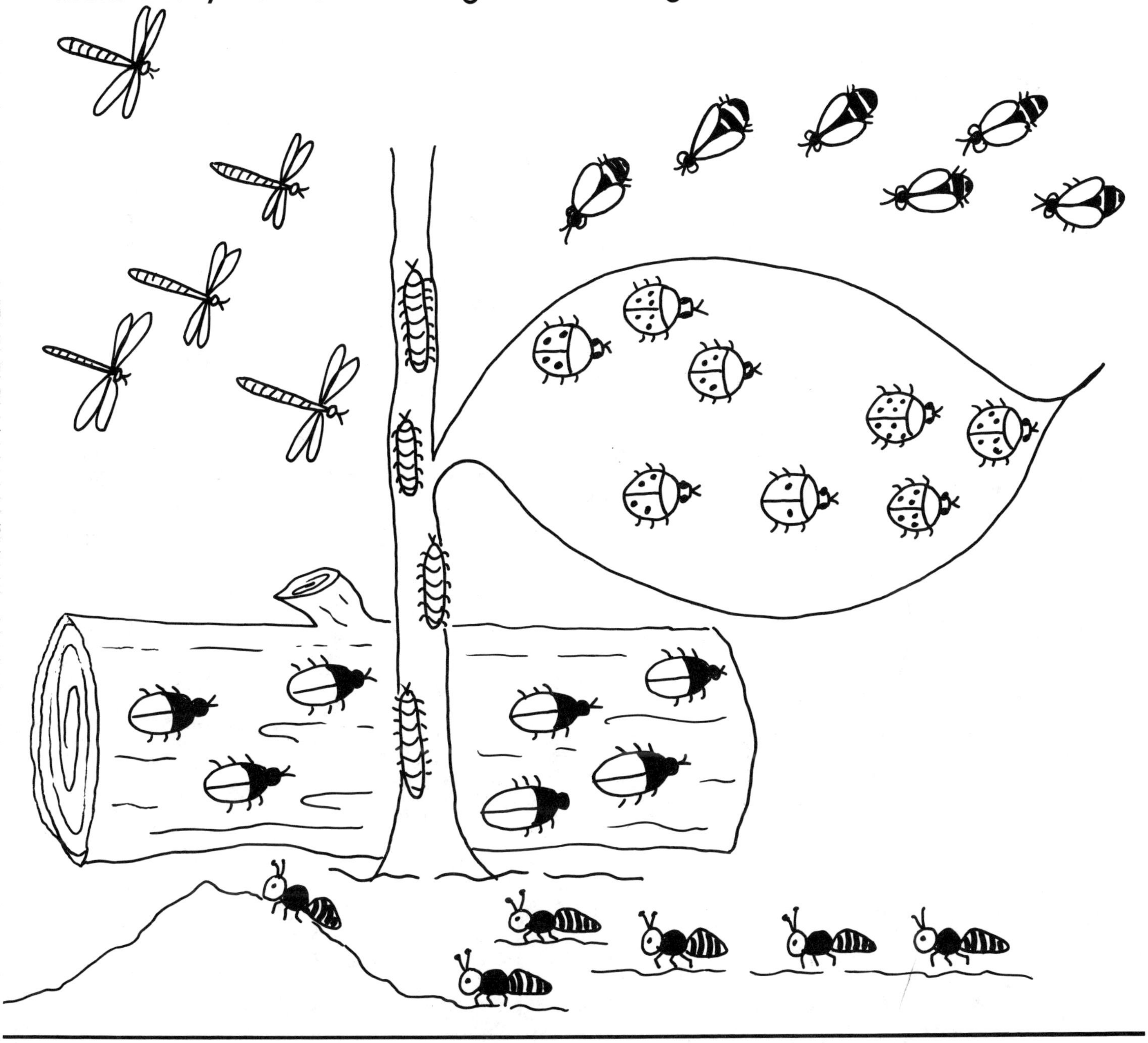

5

1989 REMEDIA PUBLICATIONS

Name _____   Tallying

Look at each person in your room. Make a tally mark on the shirt to show each shirt color.

How many? Write the number. Color the shirts.

red _____   blue _____   green _____   white _____

yellow _____   pink _____   black _____   other _____

Name _____   Histograms

Make an X to show how many beads on each string. Color the beads.

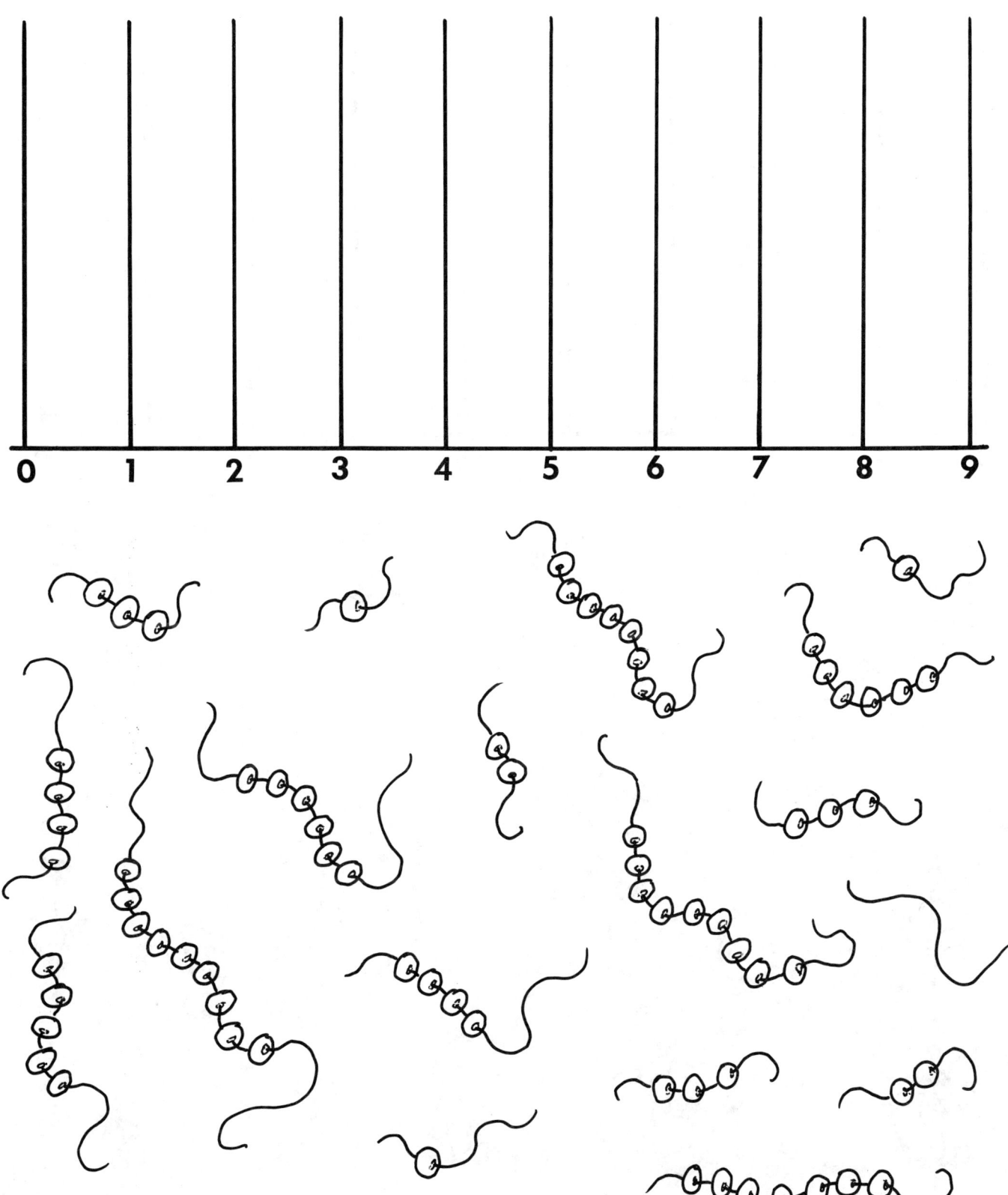

Name _____   Histograms

Make an X to show how many petals on each flower. Color the flowers.

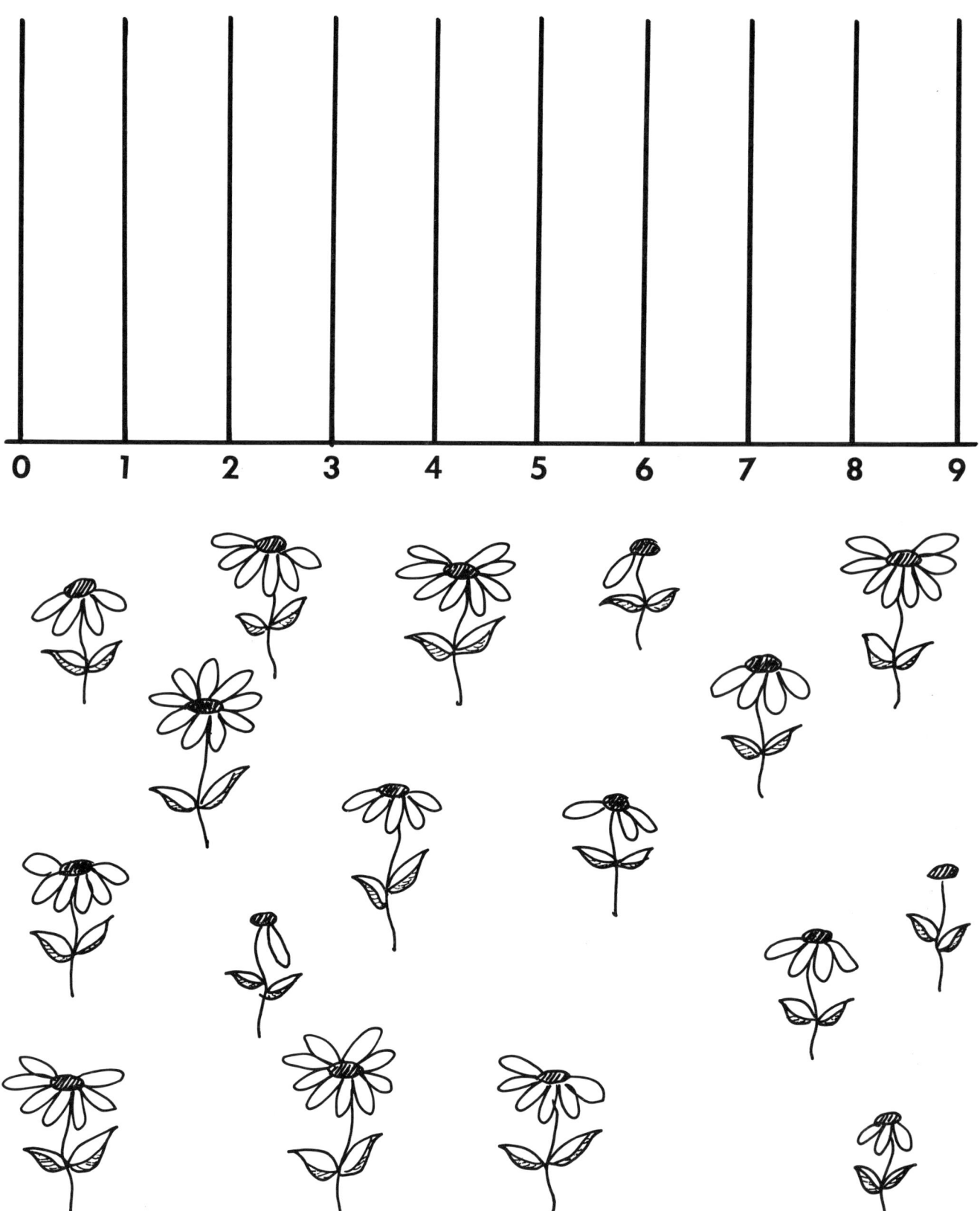

Name _____   **Histograms**

Make an X to show how many letters are in the first name of each person in your class.

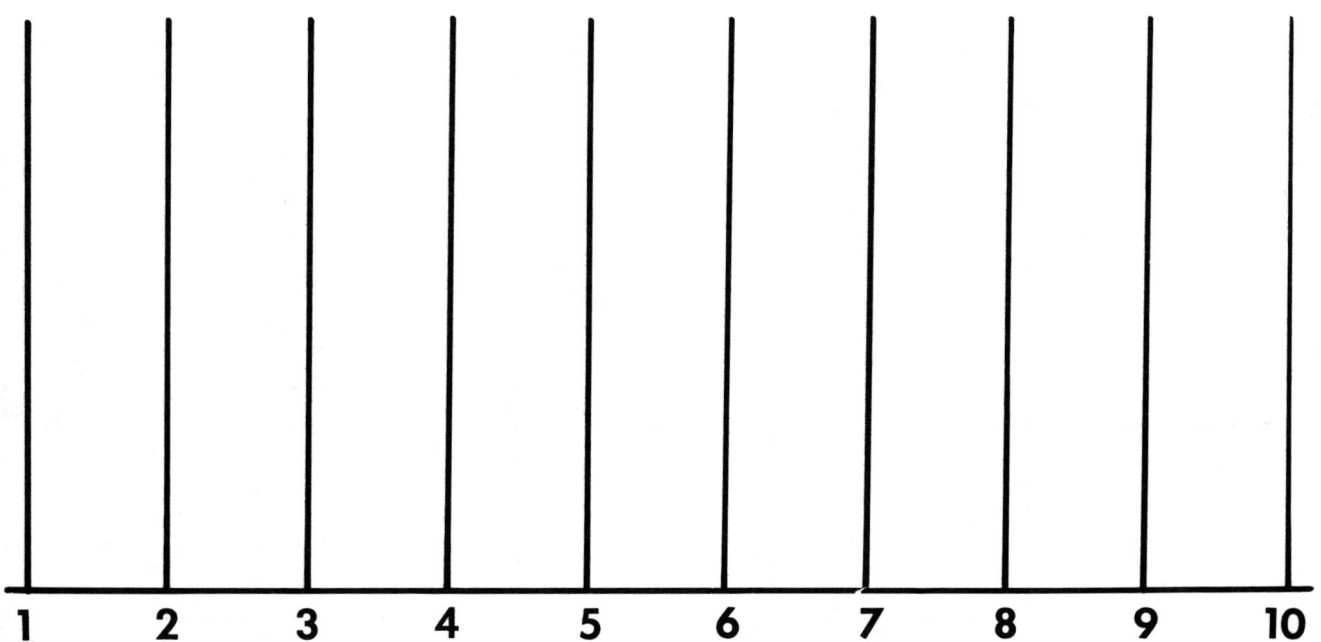

Write your name. _____

How many letters? _____

1. What is the fewest letters? _____

2. What is the most letters? _____

3. How many names had six letters? _____

4. What does each X mean? _____
   _____

5. How many names had three letters? _____

Name _____   Bar Graphs - Vertical

Cut out the animals on page 11. Glue them in the spaces. Color the animals.

 _____    _____

How many?

 _____    _____    _____

Name _____   **Bar Graphs - Vertical**

Cut out the animals and glue them on the graph.

Name _____   **Bar Graphs - Vertical**

Dan took all the coins out of his bank. Cut out the coins on page 13 and glue them in the spaces. Color the coins.

How many?

1¢ _____   5¢ _____   10¢ _____   25¢ _____

Name _____　　　　　　　**Bar Graphs - Vertical**

Cut out the coins and glue them on the graph.

Name _____

Bar Graphs - Vertical

This is Fluffy, the Prize Pig.

Fluffy has won many prizes!

Cut out the prizes on page 15.

Glue them in the spaces.

1st **2nd** **3rd** **4th** **5th**

How many?

1st _____ 2nd _____ 3rd _____ 4th _____ 5th _____

How many in all? _____

Name _____   **Bar Graphs - Vertical**

Cut out the prizes and glue them on the graph.

Name _____

Bar Graphs - Vertical

Color a space for each candy. Color the candy.

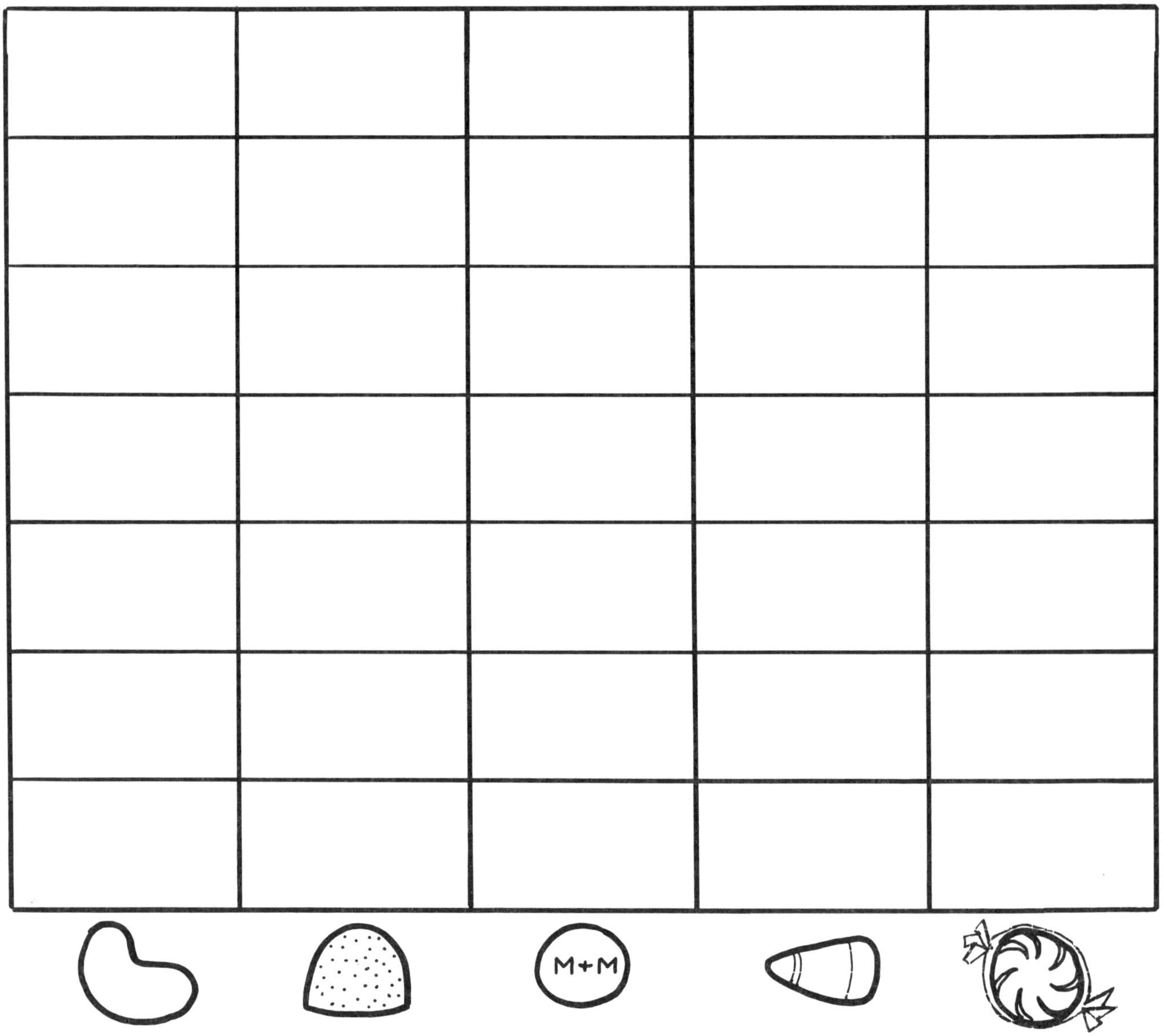

Name _____ Bar Graphs - Vertical

Color a space for each shape. Color the shapes.

Name _____   **Bar Graphs - Vertical**

Color a space for each button. Color the buttons.

Name _____   Bar Graphs - Vertical

20 children picked their favorite foods. The graph shows what they picked.

Our Favorite Foods

| spaghetti | hamburger | pizza | hot dog | chicken |

How many picked:

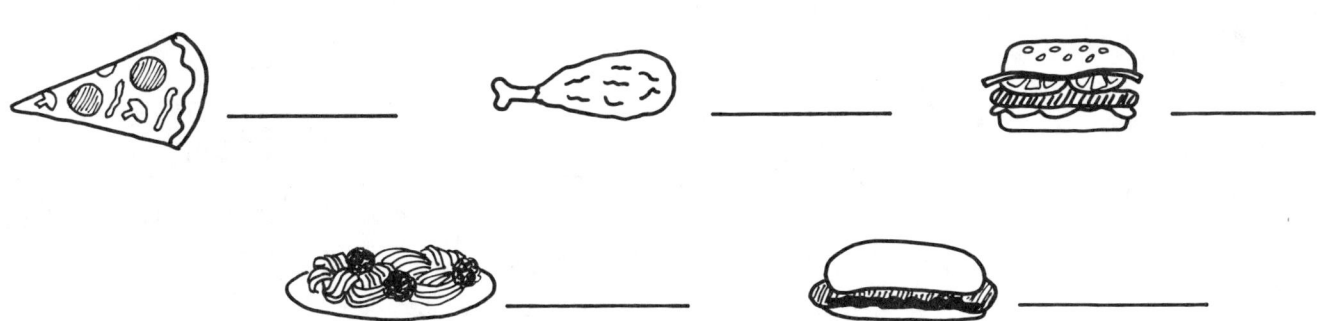

What was picked most? _____

What would you pick? _____

Name _____   Bar Graphs - Vertical

The graph shows how many of each sea animal are in the aquarium on page 21.

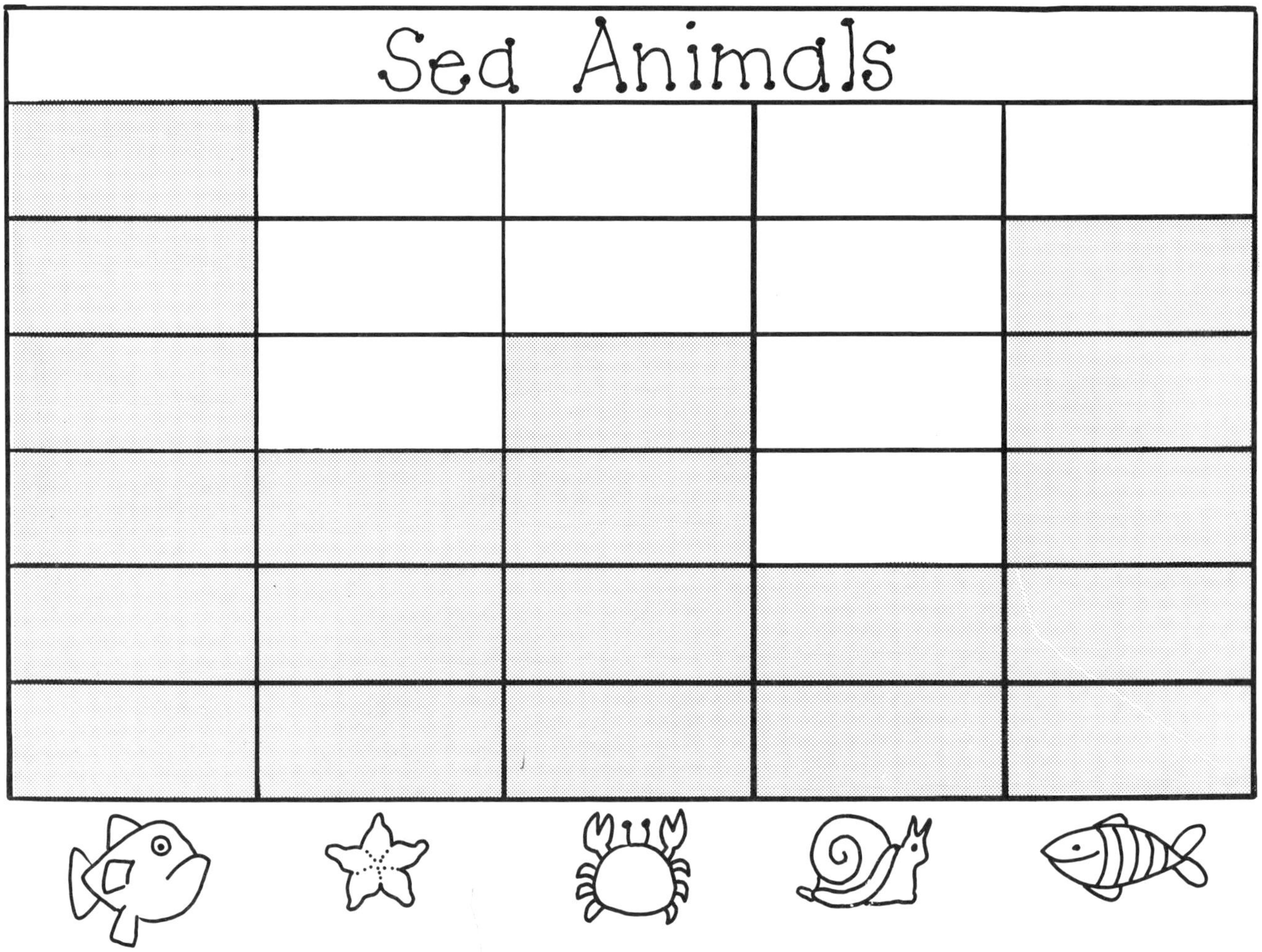

Sea Animals

🐟	⭐	🦀	🐌	🐠

How many?

🐠 ____   🦀 ____   🐟 ____

⭐ ____   🐌 ____

Draw the sea animals in the aquarium. Make the number the graph shows.

20   1989 REMEDIA PUBLICATIONS

Name _____

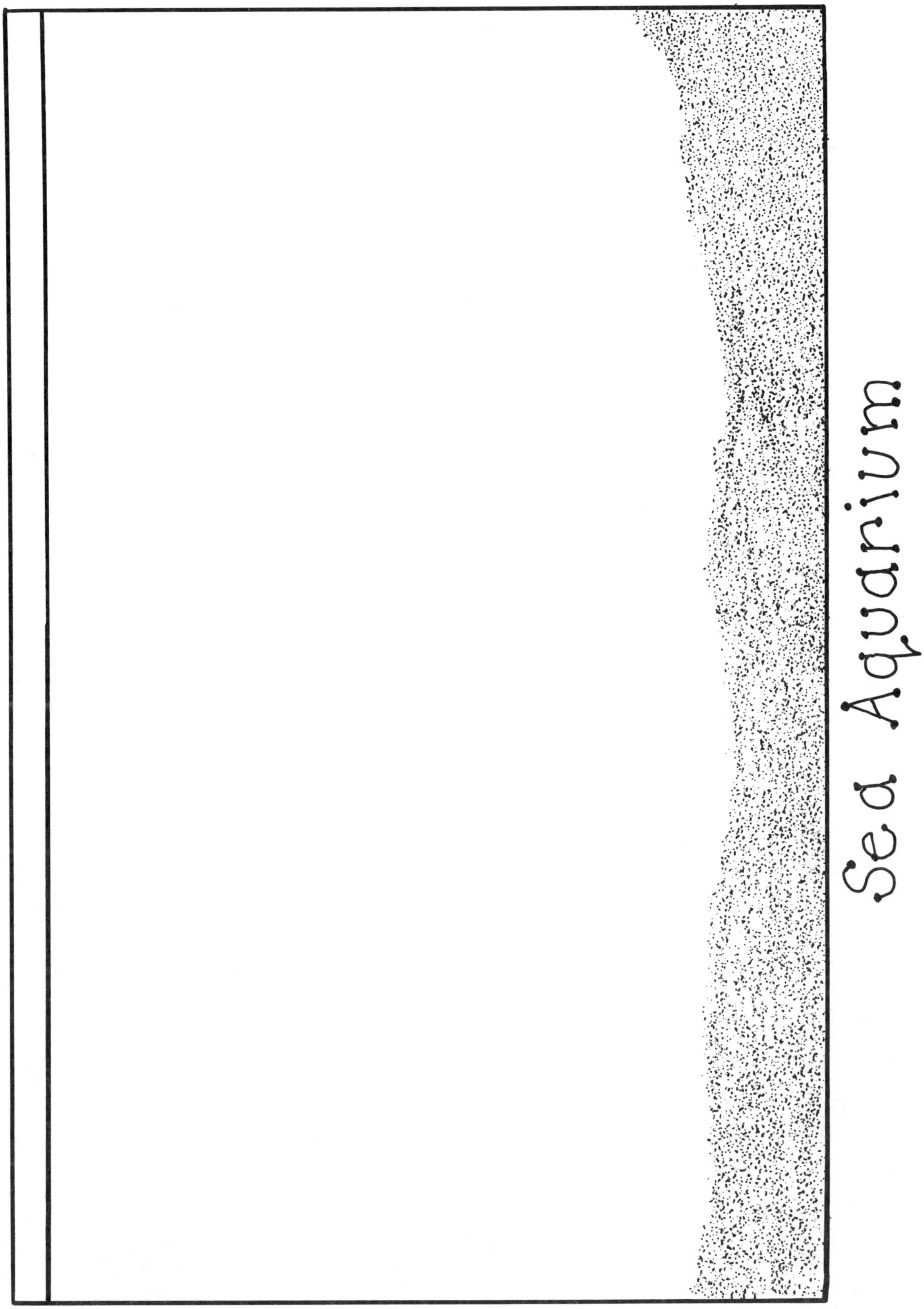

Bar Graphs - Vertical

Sea Aquarium

21

1989 REMEDIA PUBLICATIONS

Name _____   Bar Graphs - Vertical

Ask each person in your class to tell their favorite holiday. Color a space for each one.

Favorite Holidays

Christmas	Hanuka	Valentine Day	Easter	Halloween	birthday

How many?

 _____    _____    _____

 _____    _____    _____

Which is the most favorite holiday? _____

Name _____

Bar Graphs - Horizontal

Cut out the vegetables on page 24 and put them in the right space on the graph.

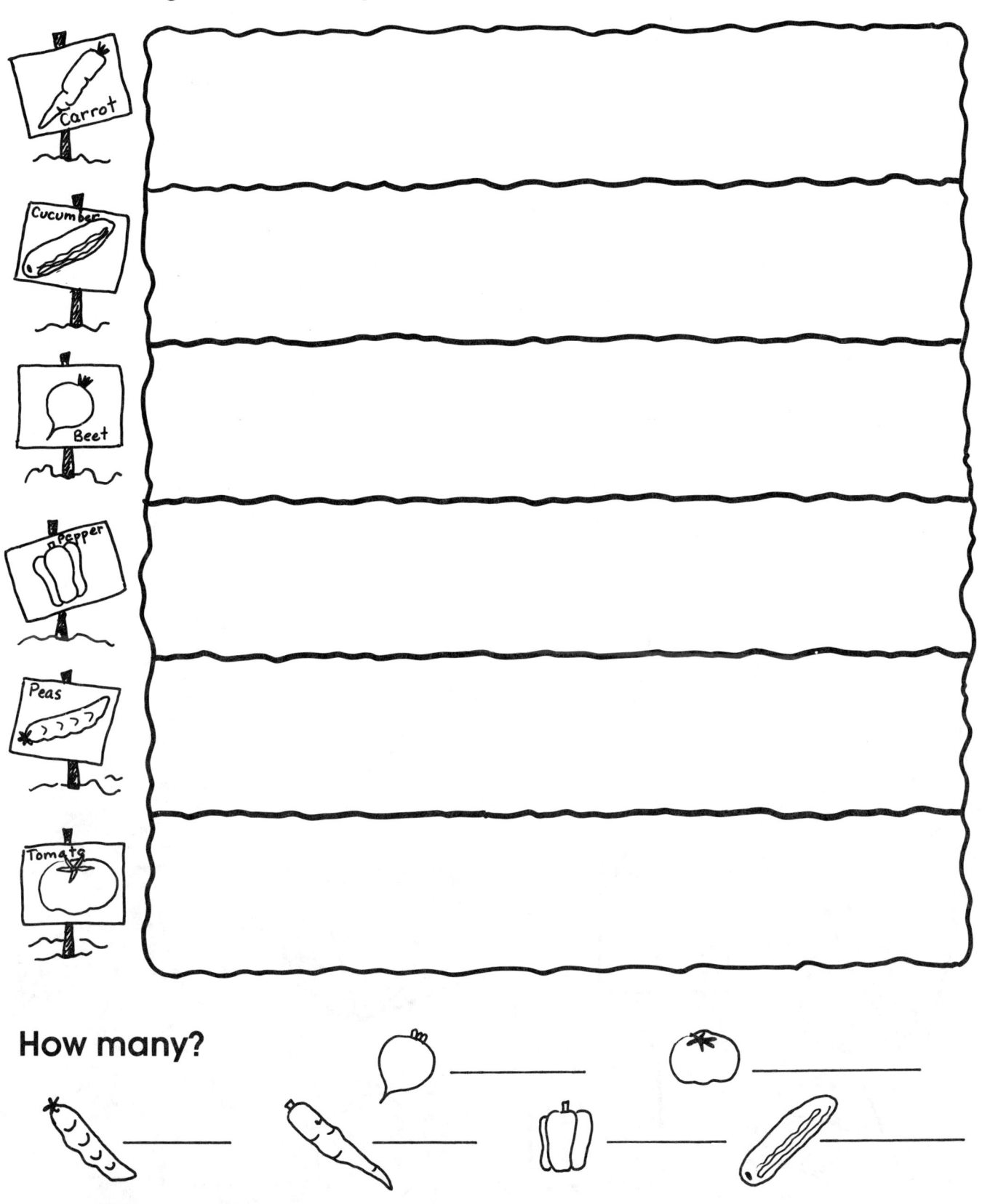

How many?

Which do you like best? _____

Name _____ Bar Graphs - Horizontal

Color the vegetables for the garden. Cut them out. Glue them in the right row on the graph.

Name _____   Bar Graphs - Horizontal

Cut out the animals on page 26. Put them in the correct row on the graph.

How many?

25

Name _____

Bar Graphs - Horizontal

The ranger is watching the animals until you put them on the graph. Color them first.

1989 REMEDIA PUBLICATIONS

Name _____   Bar Graphs - Horizontal

Cut out the shells on page 28. Glue them in the correct row on the graph.

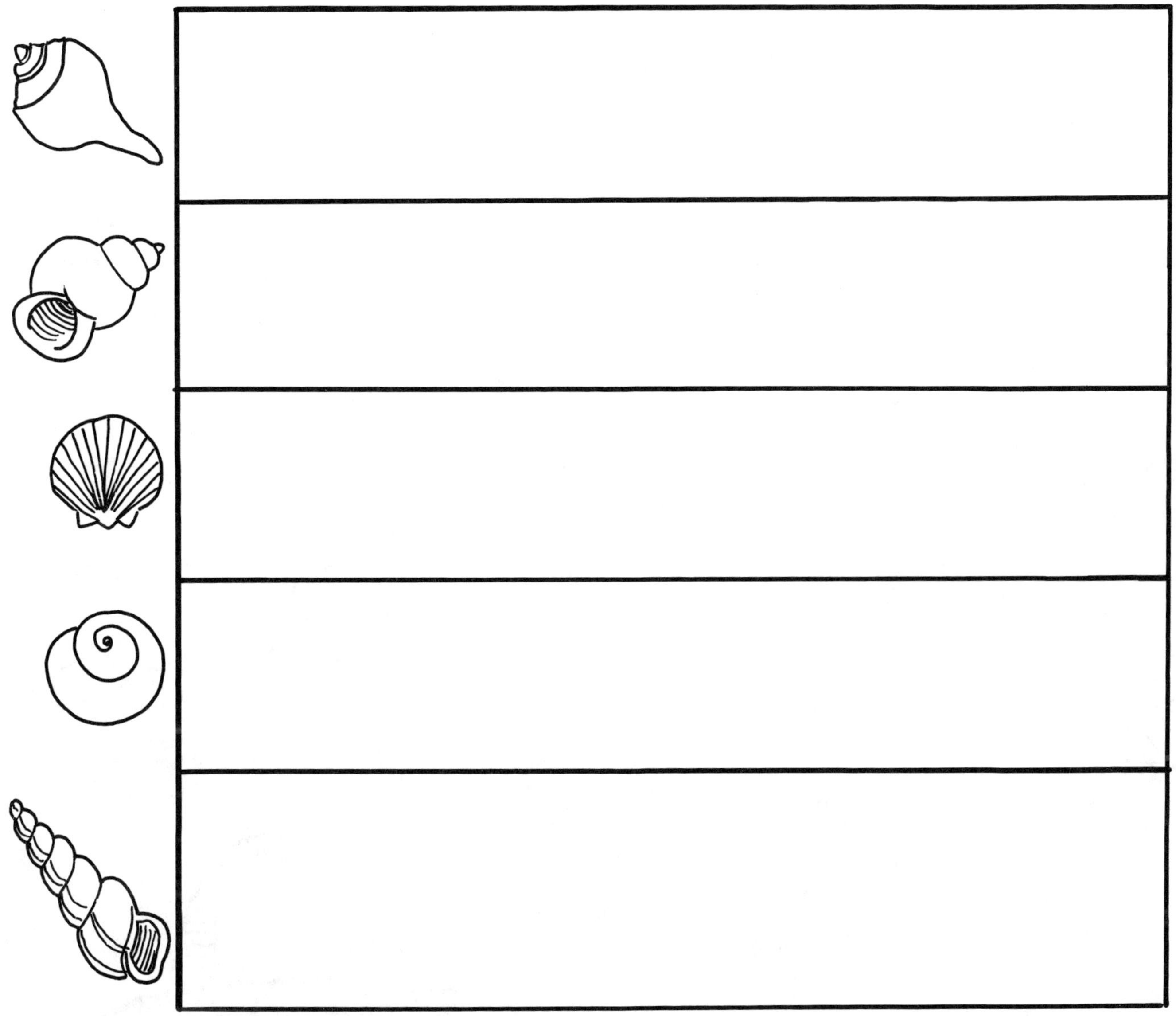

How many?

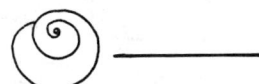 _____   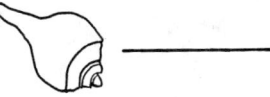 _____   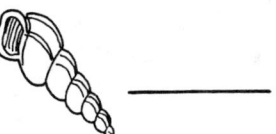 _____

Color the shells.   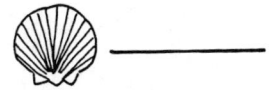 _____   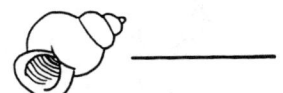 _____

Name _____   Bar Graphs - Horizontal

Cut out the shells. Glue them on the graph.

Name _____   **Bar Graphs - Horizontal**

Color spaces on the graph to show how many times each number was rolled on the dice.

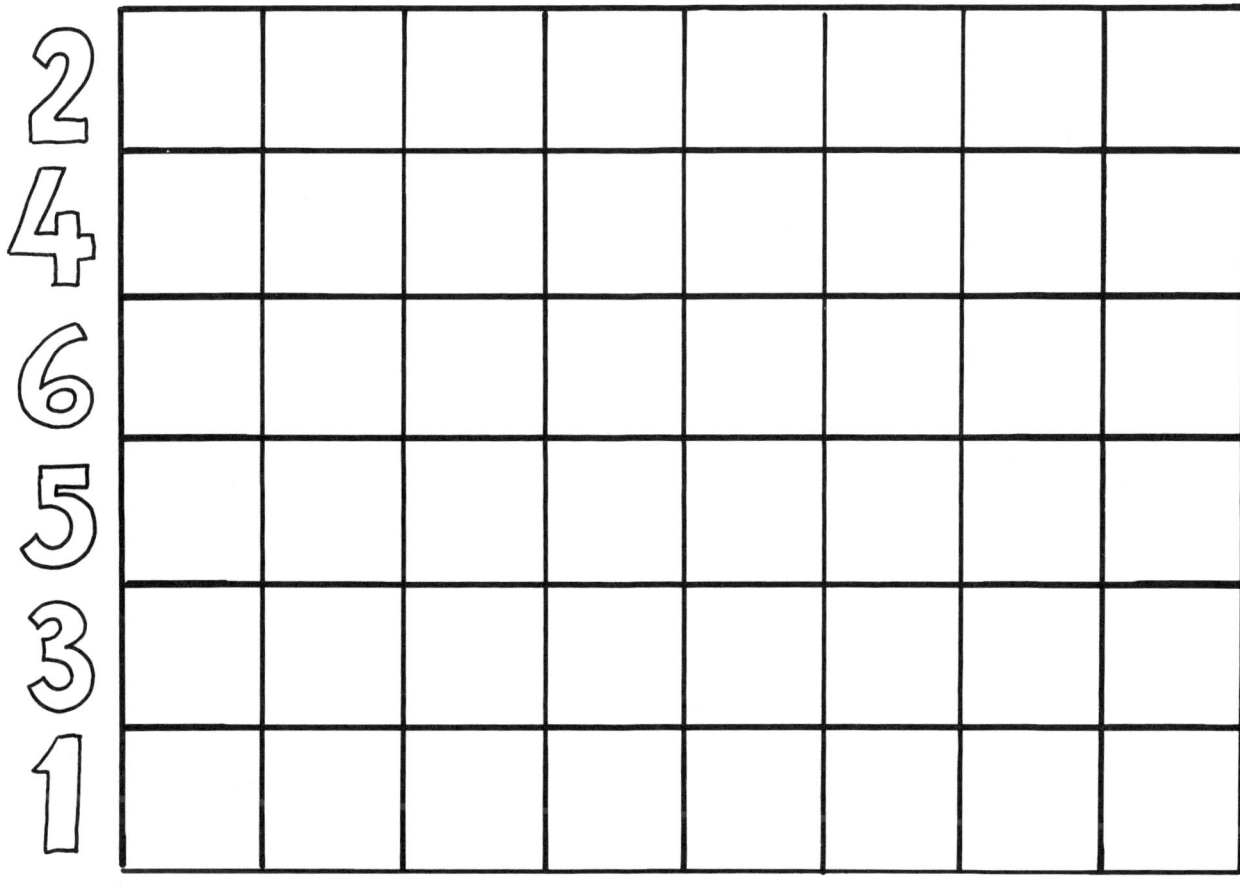

Name _____

Bar Graphs - Horizontal

Color a space for each toy on the shelf.

books

cars

balls

animals

games

Name _____   Bar Graphs - Horizontal

Monday							
Tuesday				
Wednesday					
Thursday								
Friday					

This chart has tally marks to show how many children bought milk each day.

Make a graph to show the same thing. Color a space for each tally mark.

Monday									
Tuesday									
Wednesday									
Thursday									
Friday									

How many on Tuesday? _____

Which day had the most? _____

More on Monday or Tuesday? _____

Name _____   **Bar Graphs - Horizontal**

Children picked a favorite season. The graph shows which they picked.

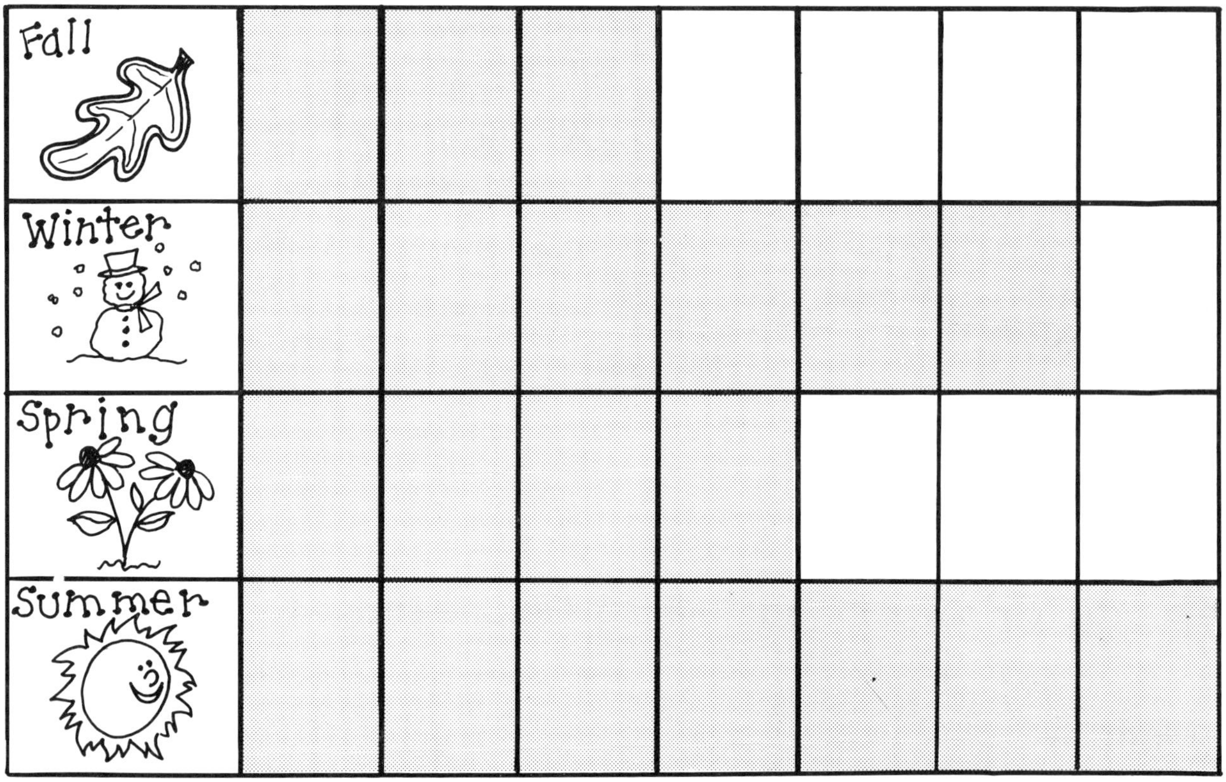

Fall							
Winter							
Spring							
Summer							

How many picked fall? _____

Which season was picked most? _____

How many more picked winter than fall? _____

How many picked summer? _____

Which was picked by four children? _____

Which would you pick? _____

Why? _____

Name _____   Bar Graphs - Horizontal

How many people in a family?
The graph shows how many.

Number of Families

Number of People in Family:
- 2: 3
- 3: 5
- 4: 8
- 5: 4
- 6: 6
- 7: 2
- more than 7: 2

How many families have five people? _____

How many do the most families have? _____

Do more families have two people or six people? _____

How many have three people? _____

Are their less with 7 or less with 6? _____

How many families have four people? _____

Color a space to show your family.

33   1989 REMEDIA PUBLICATIONS

Name _____   **Bar Graphs - Horizontal**

Ask each person in your class his/her favorite way to eat a potato. Color spaces on the graph to show their choices.

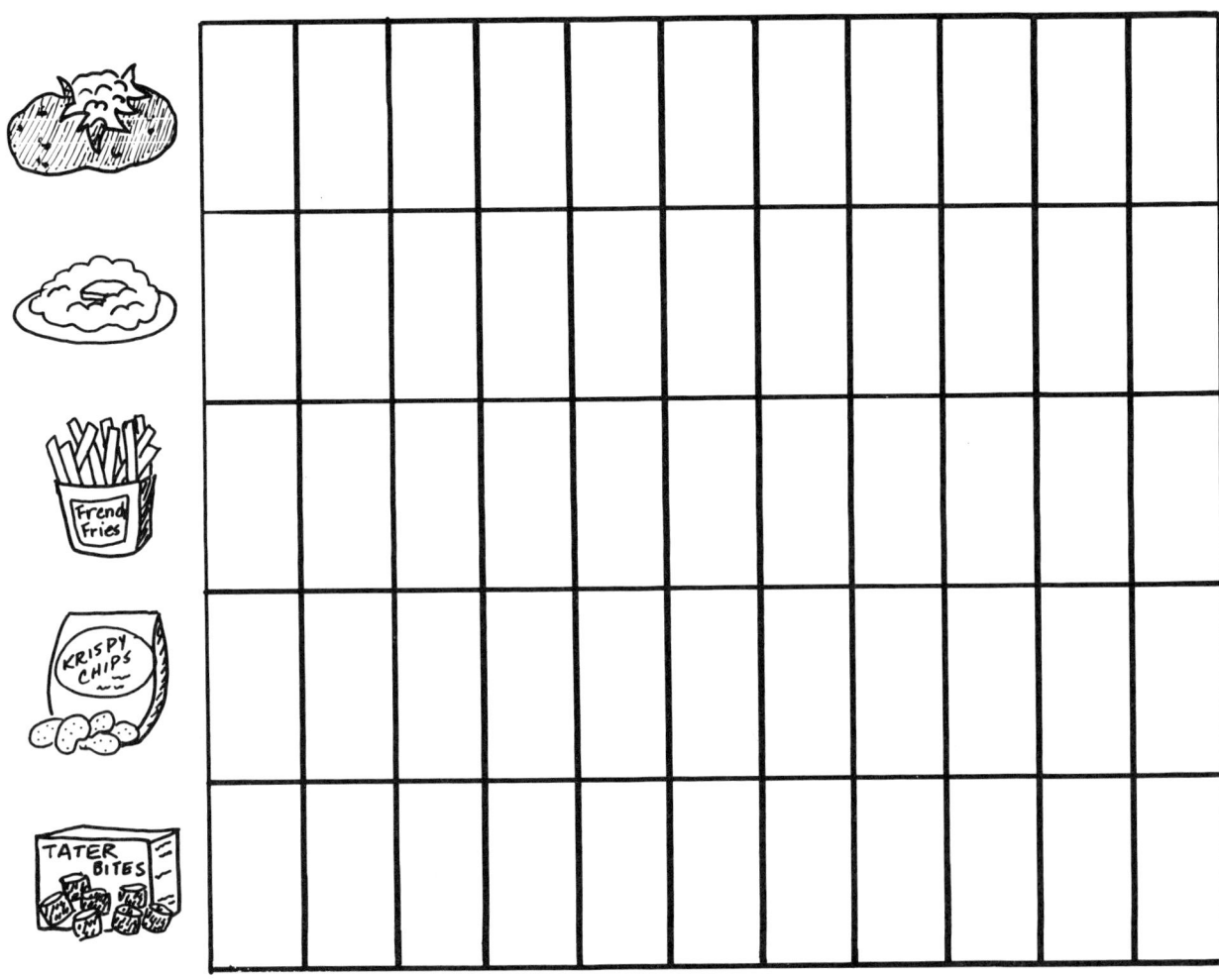

Which was chosen most often?
_____

How many picked french fries? _____

How many picked mashed? _____

Which was chosen by the fewest children?
_____

Name _____

Picture Graphs

This picture graph shows how many children are in each room.

Room 7	🚶🚶🚶🚶🚶🚶🚶🚶🚶🚶🚶🚶
Room 8	🚶🚶🚶🚶🚶🚶🚶🚶🚶🚶🚶
Room 9	🚶🚶🚶🚶🚶🚶🚶🚶🚶🚶🚶🚶🚶🚶🚶🚶
Room 10	🚶🚶🚶🚶🚶🚶🚶🚶🚶
Room 11	🚶🚶🚶🚶🚶🚶🚶🚶🚶🚶🚶🚶🚶🚶
Room 12	

There are 15 children in Room 12. Make the pictures on the graph.

How many in Room 8? _____

Which room has 14? _____

How many in Room 7? _____

Are there more in Room 8 or Room 11? _____

Which room has the most? _____

Name _____   Picture Graphs

This tally shows the favorite ice cream flavors of 25 children.

ice cream	
chocolate	𝍫𝍫 III
vanilla	𝍫𝍫
strawberry	II
bubble gum	𝍫𝍫 I
chocolate chip	IIII

Make a picture graph to show the same thing. Draw a  for each tally mark.

chocolate	
vanilla	
strawberry	
bubble gum	
chocolate chip	

36   1989 REMEDIA PUBLICATIONS

Name _____   **Picture Graphs**

Lots of cars are made every day. This graph shows how many cars of different colors were made in a week.

★ Each car on the graph means **100 cars** were made.

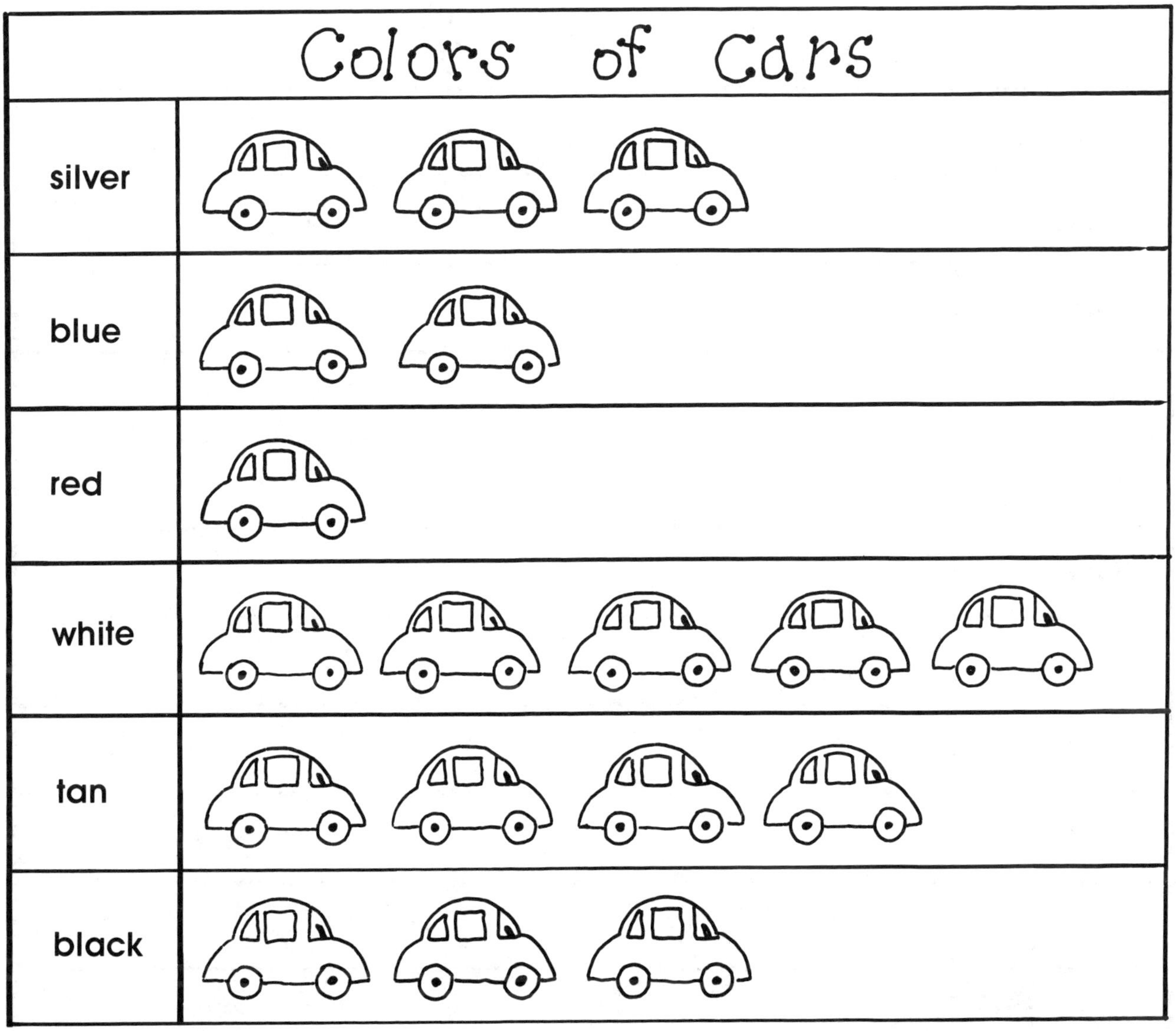

Colors of Cars

silver	🚗 🚗 🚗
blue	🚗 🚗
red	🚗
white	🚗 🚗 🚗 🚗 🚗
tan	🚗 🚗 🚗 🚗
black	🚗 🚗 🚗

How many red cars were made? _____

Were more silver cars or more tan cars made? _____

Which color was made most? _____

How many white cars were made? _____

Of which color were 100 cars made? _____

Which color would you choose? _____

37   1989 REMEDIA PUBLICATIONS

Name _____   **Picture Graphs**

Ask each person in your class the month of his/her birthday. On the picture graph, make a birthday cake 🎂 to show each answer.

Our Birthdays

January	
February	
March	
April	
May	
June	
July	
August	
September	
October	
November	
December	

What month has the most birthdays? _____

How many birthdays are in March? _____

How many birthdays are in October? _____

Color your birthday cake.

38   1989 REMEDIA PUBLICATIONS

Name _____

Line Graphs

Each week Room 8 has a spelling test. There are 14 words on each test. The graph shows how many words Ann got correct for eight weeks.

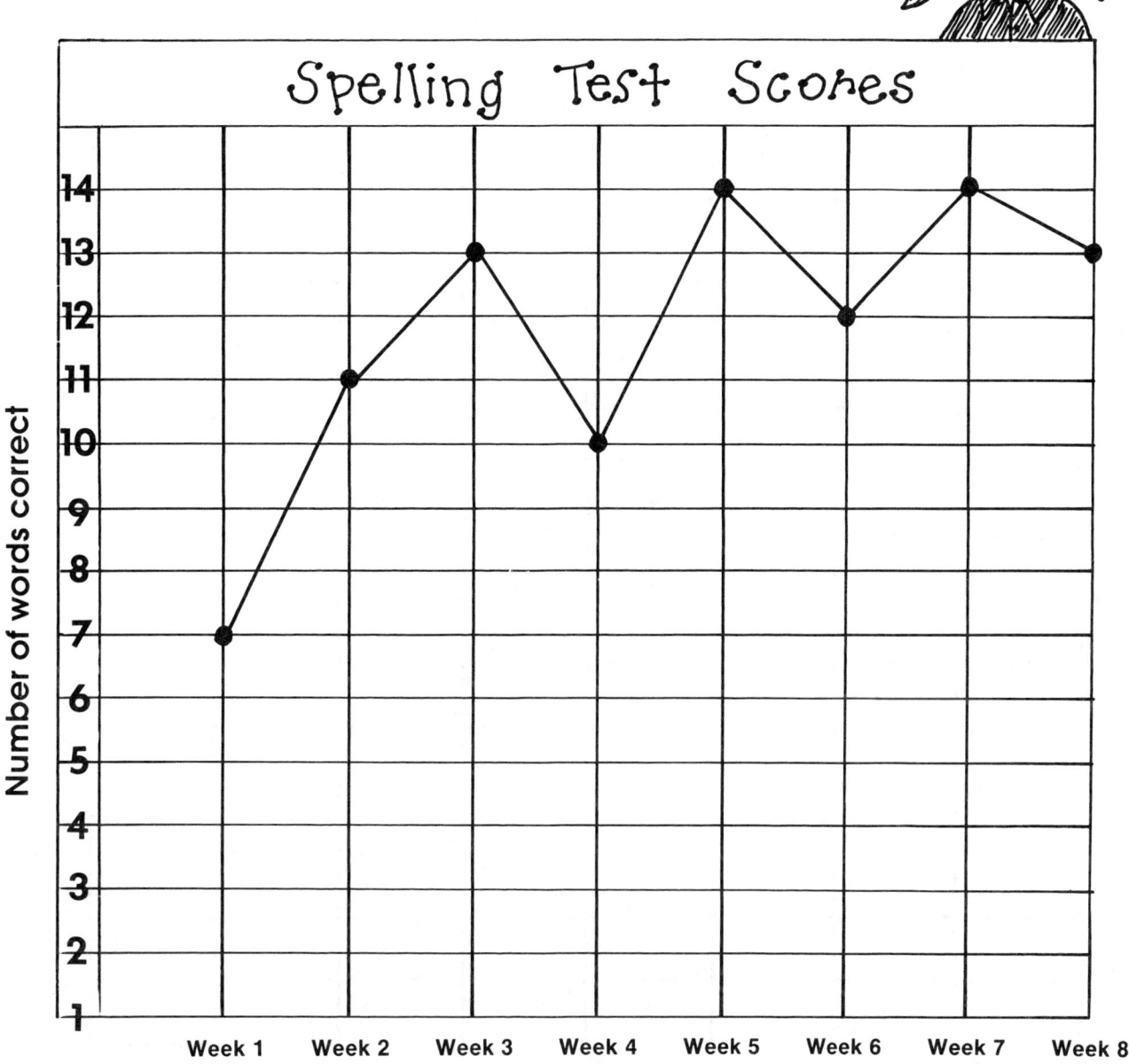

What were the scores?

Week 5 _____   Week 2 _____   Week 7 _____   Week 3 _____

Week 1 _____   Week 8 _____   Week 4 _____   Week 6 _____

Name _____   Line Graphs

The library had a reading contest. The graph shows how many books each class read.

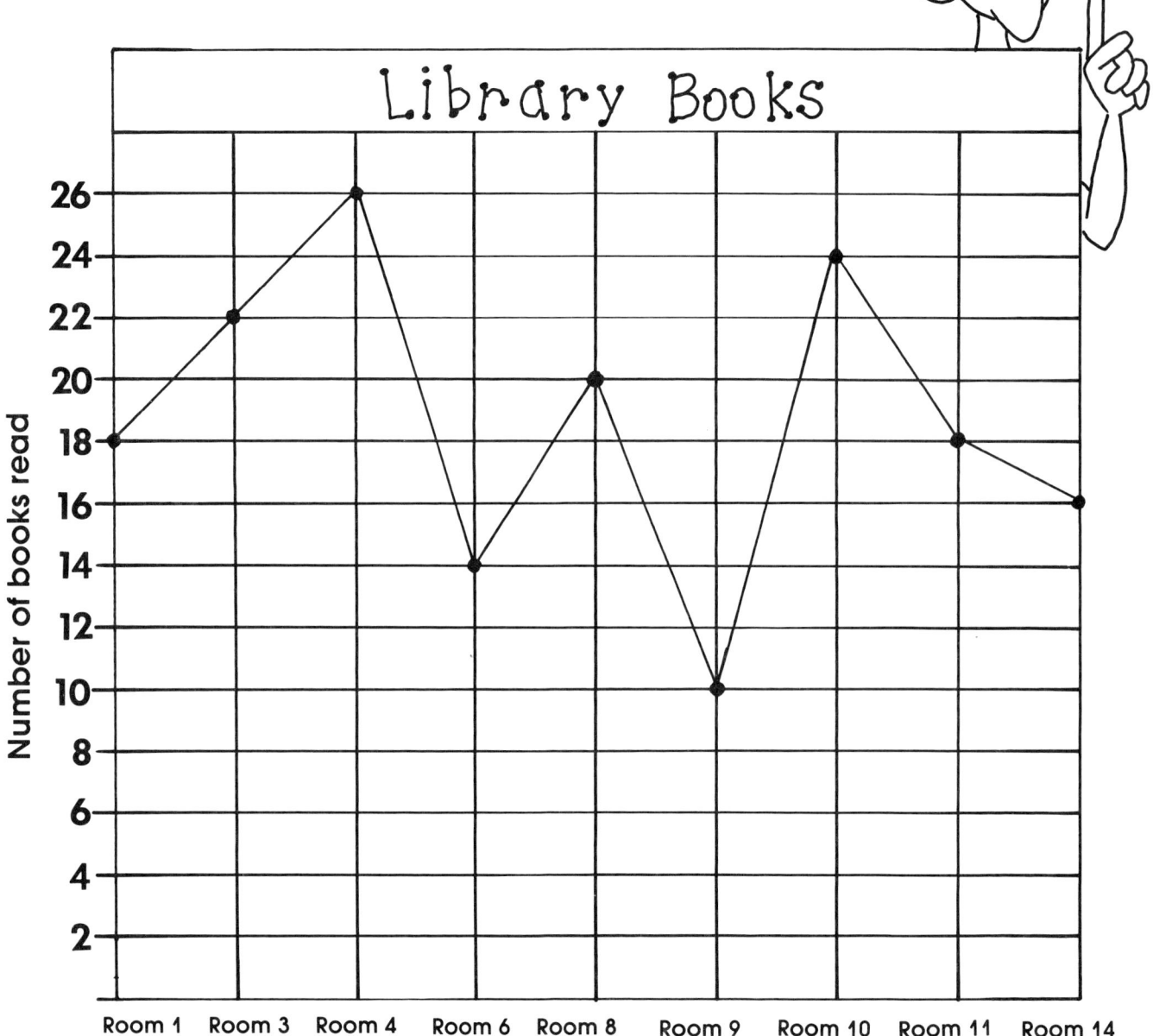

How many books did Room 10 read? _____

Which rooms read 18 books? _____

Which room read the most books? _____

Which read the fewest? _____

Name _____   Line Graphs

Use the graph to show how many children are not at school for a week. Each day, make a dot by the number that shows how many children are absent. Draw a line to connect the dots.

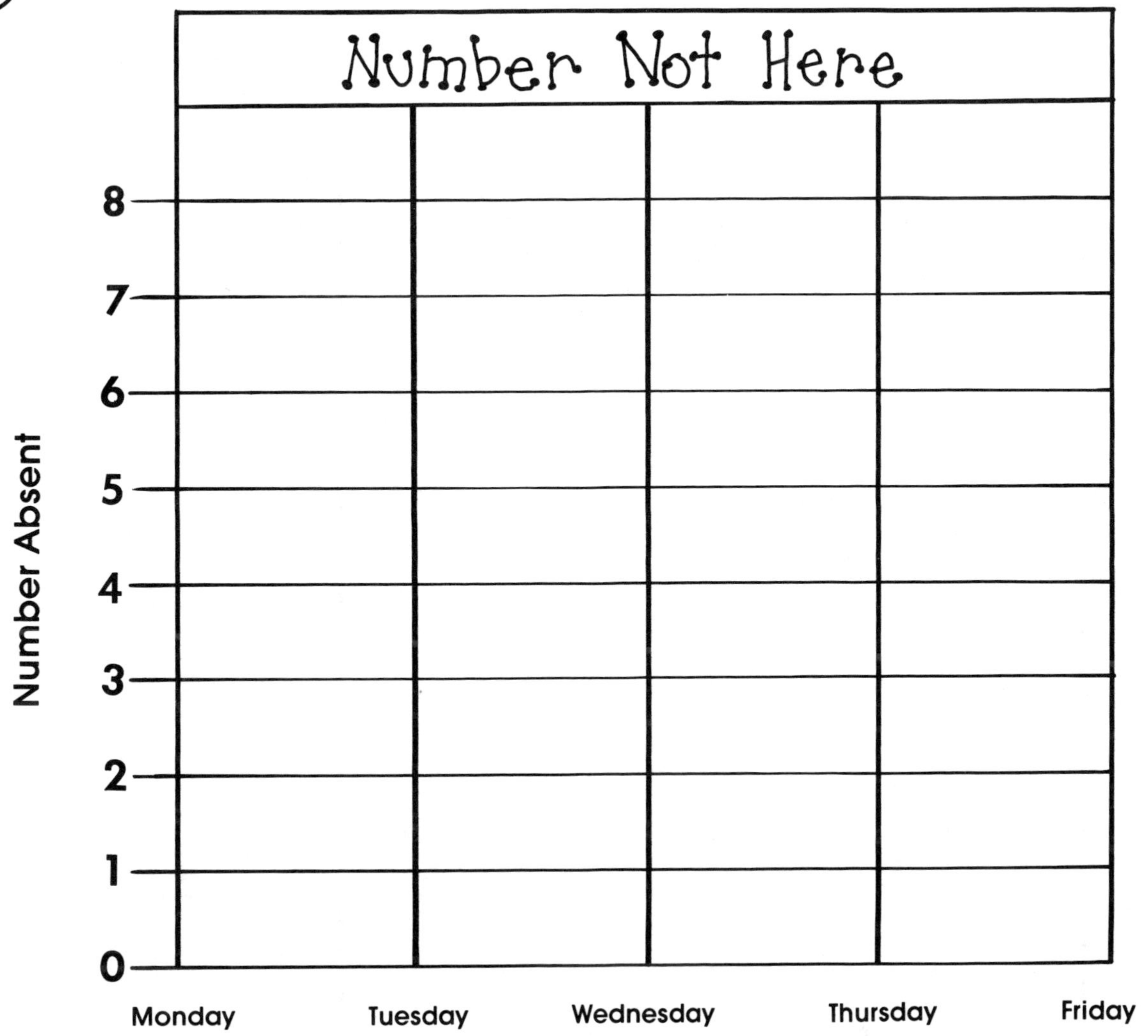

Number Not Here

(Number Absent: 0–8, days: Monday, Tuesday, Wednesday, Thursday, Friday)

How many were not here?

Monday _____   Tuesday _____   Wednesday _____

Thursday _____   Friday _____

On which day were the most children not here? _____

On which day were the fewest children not here? _____

Name _____

Bar Graph - Vertical

Name _____   **Bar Graph - Horizontal**

Name _____

Picture Graph

Name _____   **Histogram**

45   1989 REMEDIA PUBLICATIONS

Name _____

Color the space to show how you did on the graphs.

How did you do?

☺ ☹ ☹   ☺ ☹ ☹

		☺	☹	☹			☺	☹	☹
HISTOGRAMS	page 7				BAR GRAPHS — HORIZONTAL	page 27			
	page 8					page 29			
	page 9					page 30			
BAR GRAPHS — VERTICAL	page 10					page 31			
	page 12					page 32			
	page 14					page 33			
	page 16					page 34			
	page 17				PICTURE GRAPHS	page 35			
	page 18					page 36			
	page 19					page 37			
	page 20					page 38			
	page 22				LINE GRAPHS	page 39			
BAR GRAPHS	page 23					page 40			
	page 25					page 41			